⚘ *This Book Belongs To* ⚘

Name :

Phone :

Address :

Project Name :

Project No :

Date :

Foreman :

Day :

Visitors

Schedule

Problems

Safety Issues

Summary Of Work

Signature :

Employee	Trade	Hours	Overtime

Equipment On Site	No. Units

Materials Delivered	No. Units	Equipment Rented	Rate

Others

Notes :

Project Name : _______________

Project No : _______________

Date : _______________

Day : _______________

Foreman : _______________

Visitors

Schedule

Problems

Safety Issues

Summary Of Work

Signature : _______________

Employee	Trade	Hours	Overtime

Equipment On Site	No. Units

Materials Delivered	No. Units	Equipment Rented	Rate

Others

Notes :

Project Name :

Foreman :

| Project No : |
| Date : |
| Day : |

| Visitors | Schedule |

| Problems | Safety Issues |

Summary Of Work

Signature :

Employee	Trade	Hours	Overtime

Equipment On Site	No. Units

Materials Delivered	No. Units	Equipment Rented	Rate

Others

Notes :

Project Name :

Foreman :

Project No :

Date :

Day :

Visitors

Schedule

Problems

Safety Issues

Summary Of Work

Signature :

Employee	Trade	Hours	Overtime

Equipment On Site	No. Units

Materials Delivered	No. Units	Equipment Rented	Rate

Others

Notes :

Project Name : _______________________________

Foreman : _______________________________

Project No :

Date :

Day :

Visitors

Schedule

Problems

Safety Issues

Summary Of Work

Signature : _______________________________

Employee	Trade	Hours	Overtime

Equipment On Site	No. Units

Materials Delivered	No. Units	Equipment Rented	Rate

Project Name :

Foreman :

Project No :

Date :

Day :

Visitors

Schedule

Problems

Safety Issues

Summary Of Work

Signature :

Employee	Trade	Hours	Overtime

Equipment On Site	No. Units

Materials Delivered	No. Units	Equipment Rented	Rate

Others

Notes :

Project Name : ....

....

Foreman : ....

| Project No : |
| Date : |
| Day : |

Visitors

Schedule

Problems

Safety Issues

Summary Of Work

Signature : ....

Employee	Trade	Hours	Overtime

Equipment On Site	No. Units

Materials Delivered	No. Units	Equipment Rented	Rate

Others

Notes :

Project Name :

Foreman :

Project No :

Date :

Day :

Visitors	Schedule

Problems	Safety Issues

Summary Of Work

Signature :

Employee	Trade	Hours	Overtime

Equipment On Site	No. Units

Materials Delivered	No. Units	Equipment Rented	Rate

Others

Notes :

Project Name : ______________________________________

____

Foreman : ____________________________________

| Project No : |
| Date : |
| Day : |

Visitors

Schedule

Problems

Safety Issues

Summary Of Work

Signature : ______________________________

Employee	Trade	Hours	Overtime

Equipment On Site	No. Units

Materials Delivered	No. Units	Equipment Rented	Rate

Project Name :

Foreman :

Project No :

Date :

Day :

Visitors

Schedule

Problems

Safety Issues

Summary Of Work

Signature :

Employee	Trade	Hours	Overtime

Equipment On Site	No. Units

Materials Delivered	No. Units	Equipment Rented	Rate

Notes :

Project Name :

Foreman :

Project No :

Date :

Day :

Visitors

Schedule

Problems

Safety Issues

Summary Of Work

Signature :

Employee	Trade	Hours	Overtime

Equipment On Site	No. Units

Materials Delivered	No. Units	Equipment Rented	Rate

Others

Notes :

Project Name : ....

....

Foreman : ....

| Project No : |
| Date : |
| Day : |

Visitors

Schedule

Problems

Safety Issues

Summary Of Work

Signature : ....

Employee	Trade	Hours	Overtime

Equipment On Site	No. Units

Materials Delivered	No. Units	Equipment Rented	Rate

Others

Notes :

Project Name :

Foreman :

| Project No : |
| Date : |
| Day : |

Visitors

Schedule

Problems

Safety Issues

Summary Of Work

Signature :

Employee	Trade	Hours	Overtime

Equipment On Site	No. Units

Materials Delivered	No. Units	Equipment Rented	Rate

Others

Notes :

Project Name :

Foreman :

Project No :

Date :

Day :

Visitors

Schedule

Problems

Safety Issues

Summary Of Work

Signature :

Employee	Trade	Hours	Overtime

Equipment On Site	No. Units

Materials Delivered	No. Units	Equipment Rented	Rate

Others

Notes :

Project Name :

Foreman :

Project No :

Date :

Day :

Visitors

Schedule

Problems

Safety Issues

Summary Of Work

Signature :

Employee	Trade	Hours	Overtime

Equipment On Site	No. Units

Materials Delivered	No. Units	Equipment Rented	Rate

Others

Notes :

Project Name : ....

....

Foreman : ....

| Project No : |
| Date : |
| Day : |

Visitors

Schedule

Problems

Safety Issues

Summary Of Work

Signature : ....

Employee	Trade	Hours	Overtime

Equipment On Site	No. Units

Materials Delivered	No. Units	Equipment Rented	Rate

Notes :

Project Name :

Foreman :

Project No :

Date :

Day :

Visitors

Schedule

Problems

Safety Issues

Summary Of Work

Signature :

Employee	Trade	Hours	Overtime

Equipment On Site	No. Units

Materials Delivered	No. Units	Equipment Rented	Rate

Others

Notes :

Project Name : ....

....

Foreman : ....

| Project No : |
| Date : |
| Day : |

Visitors

Schedule

Problems

Safety Issues

Summary Of Work

Signature : ....

Employee	Trade	Hours	Overtime

Equipment On Site	No. Units

Materials Delivered	No. Units	Equipment Rented	Rate

Others

Notes :

Project Name :

Foreman :

Project No :

Date :

Day :

Visitors

Schedule

Problems

Safety Issues

Summary Of Work

Signature :

Employee	Trade	Hours	Overtime

Equipment On Site	No. Units

Materials Delivered	No. Units	Equipment Rented	Rate

<table><tr><td>Others</td></tr></table>

Notes :

Project Name : ____________________

Foreman : ____________________

Project No : ____________________
Date : ____________________
Day : ____________________

Visitors	Schedule

Problems	Safety Issues

Summary Of Work

Signature : ____________________

Employee	Trade	Hours	Overtime

Equipment On Site	No. Units

Materials Delivered	No. Units	Equipment Rented	Rate

Others

Notes :

Project Name :

Foreman :

Project No :

Date :

Day :

Visitors

Schedule

Problems

Safety Issues

Summary Of Work

Signature :

Employee	Trade	Hours	Overtime

Equipment On Site	No. Units

Materials Delivered	No. Units	Equipment Rented	Rate

Others

Notes :

Project Name :

Foreman :

Project No :

Date :

Day :

Visitors	Schedule

Problems	Safety Issues

Summary Of Work

Signature :

Employee	Trade	Hours	Overtime

Equipment On Site	No. Units

Materials Delivered	No. Units	Equipment Rented	Rate

<table>
<tr><td align="center">Others</td></tr>
</table>

Notes :

Project Name :

Foreman :

Project No :

Date :

Day :

Visitors

Schedule

Problems

Safety Issues

Summary Of Work

Signature :

Employee	Trade	Hours	Overtime

Equipment On Site	No. Units

Materials Delivered	No. Units	Equipment Rented	Rate

Others

Notes :

Project Name :

Foreman :

Project No :

Date :

Day :

Visitors

Schedule

Problems

Safety Issues

Summary Of Work

Signature :

Employee	Trade	Hours	Overtime

Equipment On Site	No. Units

Materials Delivered	No. Units	Equipment Rented	Rate

Others

Notes :

Project Name :

Foreman :

Project No :

Date :

Day :

Visitors	Schedule

Problems	Safety Issues

Summary Of Work

Signature :

Employee	Trade	Hours	Overtime

Equipment On Site	No. Units

Materials Delivered	No. Units	Equipment Rented	Rate

Others

Notes :

Project Name :

Foreman :

Project No :

Date :

Day :

Visitors

Schedule

Problems

Safety Issues

Summary Of Work

Signature :

Employee	Trade	Hours	Overtime

Equipment On Site	No. Units

Materials Delivered	No. Units	Equipment Rented	Rate

Others

Notes :

Project Name : | Project No :
Date :
Foreman : | Day :

Visitors

Schedule

Problems

Safety Issues

Summary Of Work

Signature :

Employee	Trade	Hours	Overtime

Equipment On Site	No. Units

Materials Delivered	No. Units	Equipment Rented	Rate

Others

Notes :

Project Name :

Foreman :

Project No :

Date :

Day :

Visitors

Schedule

Problems

Safety Issues

Summary Of Work

Signature :

Employee	Trade	Hours	Overtime

Equipment On Site	No. Units

Materials Delivered	No. Units	Equipment Rented	Rate

Others

Notes :

Project Name :

Foreman :

Project No :

Date :

Day :

Visitors	Schedule

Problems	Safety Issues

Summary Of Work

Signature :

Employee	Trade	Hours	Overtime

Equipment On Site	No. Units

Materials Delivered	No. Units	Equipment Rented	Rate

Others

Notes :

Project Name : ....

....

Foreman : ....

| Project No : |
| Date : |
| Day : |

Visitors

Schedule

Problems

Safety Issues

Summary Of Work

Signature : ....

Employee	Trade	Hours	Overtime

Equipment On Site	No. Units

Materials Delivered	No. Units	Equipment Rented	Rate